Inspired by God

Forgiven

THE JOURNEY OF A.W.B

Reverend Dr. Anthony A. W. Buie

TRILOGY CHRISTIAN PUBLISHERS
Tustin, CA

Trilogy Christian Publishers
A Wholly Owned Subsidary of Trinity Broadcasting Network
2442 Michelle Drive
Tustin, CA 92780

Inspired by God - Forgiven - The Journey of A.W.B.

Scripture quotations marked (KJV) taken from The Holy Bible, King James Version. Cambridge Edition: 1769.

For information, address Trilogy Christian Publishing

Rights Department, 2442 Michelle Drive, Tustin, CA 92780.

Trilogy Christian Publishing/ TBN and colophon are trademarks of Trinity Broadcasting Network.

For information about special discounts for bulk purchases, please contact Trilogy Christian Publishing.

Trilogy Disclaimer: The views and content expressed in this book are those of the author and may not necessarily reflect the views and doctrine of Trilogy Christian Publishing or the Trinity Broadcasting Network.

10 9 8 7 6 5 4 3 2 1

Library of Congress Cataloging-in-Publication Data is available.

ISBN 979-9-89041-991-0

ISBN (ebook) 979-9-89041-992-7 (ebook)

Dedication

To my wife, Louise Kathue Martin Buie, and to my late wife, Jeanette Elizabeth Buie, who have supported me in all my endeavors.

Acknowledgments

I give all the glory and honor to my Lord and Savior Jesus Christ. He has kept me throughout my entire life and brought me this far by His grace and forgiveness.

To my children, Leslie Lynn Baber, James Whitfield Buie, Trina Elizabeth Richards, and Jena Antoinette Drummond. I love you forever and I am honored to be your father.

Thank you to my long-time friends, Minister Jeff Howell, Deacon Darnell Samuel, Minister Sheila Belle, and Pastor Robert Garcia, for your support and contributions to this book.

Contents

Foreword

Jeff Howe

During my military career, in the 80's I met a very charismatic and creative individual whose personality was not only infectious, but powerfully confident. Anthony (A.W.) Buie was a man with initiative who was able to focus on opportunities that would produce solutions. A.W. loved DJ music, loved life, and was driven to create an atmosphere that was filled with excitement. A.W's life journey did not always exemplify these characteristics.

From childhood days in Durham, North Carolina a very full small project house with six kids, loving parents, Anthony Whitfield Buie, A.W. reminisced on the simple childhood experiences that mattered most.

This tapestry of life intersects continuously with trials, tragedies, and loss. The reader will be able to capture the "loving thread of grace" through A.W.'s life events until eventually, in 1999, A.W.'s life search collides with the loving, passionate, merciful God the Father, with whom he fell in love and is "oh so passionate" today to proclaim the good news of the Gospel of Christ. New faith, new fire, is ignited inside of A.W., stirring life-long gifts and watching how God was redirecting his thinking, his passions, and his desires.

From Durham, NC to Fairbanks, AK to Louisville, KY, the reader that embarks on this "A.W." journey will experience a transformation that is fueled by the love for God, and for this country. A.W. seeks to proclaim the very essence of the Father who forgives, never leaves, nor forsakes, and always provides!

A vibrant ministry that will infiltrate the fiber of your being and reach the heart of our nation, one can quickly recognize Anthony Whitfield Buies' journey is *Inspired by God*.

"Focus on what can be done through God as opposed to what cannot be accomplished by man."

Jeff Howe
Creative Director and Operations
"Together We Stand"

Foreword

Dr. Robert S. Garcia

My wife, Gretchen, and I had the joy and blessings of serving as Interim Pastor at True Victory Baptist Church in North Pole, Alaska from 2008-2009. We were assigned under the American Baptist Churches for a one-year service; however, we stayed an additional month at the church. It was very diffcult to leave the congregation of True Victory. Our experience ministering in an African American church was a blessing beyond words. The encouragement to preach and teach God's word was amazing. How this congregation loved God's word!

It was then that we had the privilege and honor of meeting A.W. Buie and his wife, Louise. The Buies kept in constant contact with us. We developed a deep gracious friendship and spiritual kinship. Gretchen and I are thrilled that our relationship with the Buies is a lasting one.

We are also honored to personally know the "changed" A.W.B. It is with the greatest respect and honor that I write this piece as an addendum to his life story. This story will challenge your honesty, faithfulness, loyalty, and unconditional love. God continues to minister through the dedicated lives of the Buies. They are genuine "Doers of the Word."

In ever-growing love and appreciation,
Sincerely,
Dr. Robert S. Garcia

Foreword

Sheilah Belle
"The Belle"

We are our choices! Every choice we make makes us who we are! From the time we are born until the time God calls us home, we are who we are because of the choices and decisions we make in life!

It was in 2004 when God gave me the opportunity to make a choice to trust Him in literally new territory. I was invited to participate in what would soon become a project that would change my life forever. The visionaries of this project, Anthony Buie and his wife Louise, created, TOGETHER WE STAND. In their first major effort to put a face to this undertaking, they put together a gospel concert. The setting was Fairbanks, Alaska and they invited me to emcee this event as well as cover their first Gospel Fest as a journalist. Held at the Carlson Center in Fairbanks, Alaska, only God knew what was being put together in this unfamiliar

Gospel music territory, as their other special guests also included the award-winning Mali Music, and now top music executive Phil Thornton.

While the climate only produced minus eleven-degree temperatures as the high each day during our entire weekend, the warmth that was extended to me and all the guests was merely a reflection of the love of God the Buies had in them.

While there, we shared, broke bread together, praised the Lord, and became friends. Anthony Buie shared some personal moments of his life with me that also changed my life. His heart for God was real and at times heavy because of so much he was able to chart his way through.

While growing up in North Carolina, Anthony experienced many challenges. From a very young age, Anthony saw not from the white community but from his own community, because of his darker pigmentation. From almost losing his life in his younger years while playing to remembering when he burned his wrist on the hot stove in his home while trying to stay warm, these baby thorns would often remind Anthony of God's mercy.

He came from a hardworking family that persevered and kept the family store going and prayed their way through any other unexpected situation in life, while attending church four to five times a week.

All that he learned while at home gave him a strong foundation to stand on after he joined the Air Force and experienced a few other life lessons that would ultimately challenge his commitment to God.

By the time I met Anthony Buie, who I now call Uncle Blu Blu, it is now easy to understand why he has never turned away from God because of so much God has brought him through. His love for God runs deep in his DNA. From North Carolina, to San Antonio, Texas, Salt Lake City, Utah, Thailand, the Philippines, Germany, Louisiana, Nebraska to Alaska, God was ordering his steps. Even with the many ups and downs in his life, Anthony was never forgotten, and God will never forget him.

What I have learned from Uncle Blu Blu over the years is priceless. Truly I am grateful for his love and the love of his wife, Louise, as they have become a part of my family!

Truly, we are our choices, and every choice we make makes us who we are!

Sheilah Belle "The Belle"
CEO The Belle Report
CEO, LaBelle & Associates

Foreword

Darnelle Samuel

Anthony (AW) Buie exposed his life to inspire one to follow Christ. He shares his story in this book as a testimony that God's purpose for one's life is in His hands. A.W.'s journey begins as a child and ventures through obstacles in the U.S. Air Force. Many are called but few are chosen. God's servants answer the call.

Retired technical Sergeant USAF
Darnell A. Samuel

Introduction

Inspired by God, the Journey of AWB is clearly the story of my life, but I have had so many other challenges, hardships, trials, tribulations, and mistakes that I couldn't put them all in this book.

My goal for this book is that you see the hand of God, His grace, mercy, forgiveness, and redemption occurring throughout my life.

I'm not perfect, never claimed to be, but through God's saving power I am fully driven and committed to living for Him. I want to be a witness for Him and lead others to Christ as long as I have breath in my body.

I trust that you'll be touched, blessed, and have a desire to know more about the God that I serve once you've finished reading my story.

AWB

CHAPTER 1

Home

Red dirt, green grass, tall trees, and bright sunshine. There was a chicken factory nearby where we lived and often the chickens would get out. My mother said, "Catch a chicken if you can," and I did. Then she would wring its neck and then we'd have dinner. That's home to me in Raleigh Durham, North Carolina. My name is Anthony "A.W." Buie. I was born in Durham, North Carolina, in 1951. My mother and father called me Pete because they didn't want me to be a junior. That's my nickname. I was the first-born son. **Jeremiah 1: 5.** My mother knew that there was something special about me when I was born. **Galatians 1:15.** She couldn't quite put her hands on it because she was on a journey herself to find the Lord and her own purpose.

We lived in the projects, and with me being the only boy and the oldest, my father and I were close. When I was about four years old, my father took me out to fly a kite in a field near the projects. My father was holding the string along with me, and we were flying the kite together. A wind came through and the string broke. I started crying. At four years old it is a normal reaction to cry, especially when you're having so much fun. I remember my father was running to get the rest of the kite that was seemingly getting away by the shortened string. I was so happy that he retrieved my kite. That was an amazing day! Something about having that time with my father and the simple thing of flying a kite is something that I've always remembered. It's the simple things, pleasures, and events in life that matter most to a child.

In the projects, we probably lived in roughly six apartments through the years, but my father was a hardworking man and he was striving to get us out of there. My father was an electrical journeyman. He did just as he said, and we moved from the projects. The apartment I remember the most was a two-bedroom apartment on Austin Avenue off of Main Street. It was small, but it wasn't the projects. With just two bedrooms, my mother and father were in one room and my sisters stayed in the other room. I was the only boy, so my bedroom was the living room. The living room had a couch and a kerosene stove. You can imagine that there wasn't much privacy, so my life was an open book to my family. Besides the two bedrooms, there was a kitchen, one bathroom, and a couple of closets for clothing and not much storage. It was what my parents could afford, so that was life in the Buie household. When my aunt, my mother's sister, was murdered, her two daughters came to live with us. The two-bedroom apartment now had six kids in it. It was crowded, and there was much to do with so many kids and now eight people in that little apartment, but we made do with what we had.

It was really something to live on Austin Avenue and to be out of the projects. It was a struggle for my mother and father, but they did it. We had a little dog called FiFi. We were able to care for and love something. We kept that dog for a while, but FiFi became older; eventually she passed away.

One day I remember sitting at the kitchen table eating breakfast. An eagle flew on the porch right in front of me. Looking at the eagle, I said, "Wow!" I asked my mother, "What was that?" She said that it was an eagle. It was amazing to see that huge eagle with its large wings and hooked beak. Up close, an eagle is still a majestic and awesome bird. That one sighting has stayed in my mind and has impacted all that I've done from then even to now. **Isaiah 40:31.**

North Durham, North Carolina

I remember one cold morning all six kids were around the stove trying to keep warm. The stove was in the living room, so that was the place to be to get warm in the 1950's. I was close to the stove with my back side next to the stove. I put my arms behind my back. I remember I was so close to the stove that I burned my wrist. The burn puffed up and as boys will do, I started playing with it with a pencil, poking it all the time. I eventually wrapped it up and put Vaseline on it. I was young and a boy, but it's amazing thinking about it now: I didn't realize that poking that spot on my arm would leave a scar or a mark, but it did. I have that mark today on my wrist where I played with that burn for a long time. My parents just let me be a typical boy, especially since they had all those girls. I'm also amazed that it took the longest time to heal, but that's what you get when you play with a sore. I didn't ask my parents how to handle it or run to them for advice. I just did what my little boy mind told me to do. I didn't run to my parents with stuff like that, but left to my own to figure out, it left a scar. **I Corinthians 13:11.**

School

I remember when I was in first grade or kindergarten, my mother took me to school. I cried because home and my mother were all I knew. My mother left me at school and told me I'd be all right; eventually I liked going to school. I found out that I really liked the girls, too. I made friends with some of the guys at school. I also went to church several days a week with my mother. However there was something about going to school that I really liked. I enjoyed being with the other kids. I got to express myself, so I talked a lot in school. I always wanted to be special, to stand out and be the leader. My favorite thing was being the line leader and leading the entire line of students right next to the teacher. The teacher often rewarded us when we cleared off the chalkboard or did other helpful tasks in the classroom. One thing about back in those days, the teacher rewarded us and she also disciplined us. She would discipline us with a paddle. Each teacher would help you to grow up in school. Teachers were part of your upbringing and guidance, and they were a rearing village of sorts.

Hebrews 12:5-11.

I attended an all-Black school. The school was the East End Black Teachers District back then. Matter of fact, I didn't have any connection to any White people at the time I was growing up at all. **Acts 10:9-22.** I liked school, but I wasn't a scholar by any means as far as my academics. Back then, like now, kids would tease you and call you names. Kids called me names like "Black Buie" or commented on how dark I was, and that really hurt my feelings. As time went on, I totally forgot all about it when James Brown and his bandleader, Alfred "Pee Wee" Ellis, came up with the song, "Say It Loud – I'm Black and I'm Proud." I was good then.

My mother worked outside of the home at Duke University. She had a job cleaning the dormitory rooms.

Once there was a big snow storm, which was unusual for Durham, but the snow came anyway. My mother told us to stay home because it was snowing outside. I disobeyed my mother and went to school anyway, in the snow. There was a highway in front of us at the apartment where we lived, and how I crossed that street as a kid, I will never know. I was in the first grade and went to school in the snow anyway. Sometimes when you're a kid you don't know exactly how much danger you are in when you take a chance and do certain things, especially things that your parents tell you not to do. At times, it seemed like her angel was hovering over us, me and my sisters, while we were going to school.

My sisters and I played outside in the dirt, as I said before. There weren't many sidewalks but a lot of dirt and playing tag back and forth. There was a park, but we were too young to go by ourselves until one day.

I remember that hot summer day. Our parents went to work, and we were supposed to stay at home, as usual. I believe I was probably about in the second or third grade; my sisters wanted to go meet some boys, I believe. We crossed the street again and went into the woods over there, past the department store, and in the woods there was a swimming hole. I jumped in the swimming hole because it was so hot. I couldn't swim, so I actually started drowning. I guess some way, somehow a person came along and picked me up and pulled me out of the water. I believe that was an angel because we didn't know the person that pulled me out of the water. I always wondered, "Who was that guy that pulled me up?" because I was gasping for air and trying to catch my breath. It was so dark and I was struggling trying not to drown. The girls and I made each other a promise that we wouldn't tell our parents, not a word. Looking back, we realized that life is so short; our aunt had already died. After that, did I learn to swim? I definitely learned how to swim.

We moved to the other side of town. I was riding my bicycle on the back side of the house. There was a moving company moving a family into the area. The truck had a loading dock or ramp on the back of it, and me being me, I thought I could go under the loading dock, this metal loading dock/ramp that went up and down. I thought I could be a risk taker and go down a hill on the street nearby and go underneath the metal loading ramp. As a daring boy, I went down the hill, and when I got close to the truck, it caught me on my head. It should have killed me and but just knocked me down. But I had a big knot on my head for a long time. I know that God spared my life one more time even though I did something so dangerous. I didn't even get a whipping for my carelessness. I can truthfully say that I deserved it that time. My mother would whip us with anything she could get her hands on, including extension cords, etc., but I should have gotten a whipping for that act.

After my aunt passed away, she left everything to my grandfather, which would have been her father. My aunt owned a house and my grandfather told my parents that we could live in the house. My auntie owned a neighborhood grocery store and my grandfather, we called him "Papa," ran the store. Getting the house meant it was time to move again, from apartments to finally our own house. So we moved from North Durham to South Durham. In the neighborhood of this new house is where I started learning how to swim. There was pool a block away that allowed Blacks in the public swimming pool. In the summertime, I went swimming every day, taking swimming lessons.

Until we moved in the house in South Durham, our parents walked everywhere until they learned how to drive. Because of my aunt, my mother and father finally learned how to drive a car. My aunt had a 1948 Roadmaster Buick that became ours. All of us fit nicely in that big car.

Church

My mother took us to church in the Roadmaster. Prior to that, we walked to church on the highway. Sometimes the saints would pick us up, but primarily we walked. Because my two cousins, Paula and Cynthia, came to live with us, we were all raised as siblings rather than cousins.

Walking to church, my father would be with us to protect us from the dangers on the highway and from anything or anybody else that would hurt us. We were primarily going to Bible School and any church service. We went to church a lot and most of the days of the week.

When we moved in this house and had a grocery store, our "Papa" became an even greater fixture in our lives, along with our mother and father. I guess because I was the boy, my grandfather took a liking to me more than the girls. He would take the time to tell me a lot of things, a lot of good stuff that a boy should know, along with some old stuff or history of our family. I was really thankful that he was able to provide that to me because of the intense struggles that I was going through around the third grade.

In third grade, I struggled with reading and writing. Back then there wasn't a lot of outside help and if you didn't get it right, it wasn't very pleasant for you in school. I wish my mother and mather had helped push and tutor me in reading and writing. They were working and providing for the family, so they didn't have the time. I don't remember getting any type of awards for academics throughout my school years. I do remember that we spent a lot of time in church, singing, going to Bible School, revivals, and having a good time in church.

Store

Church was a huge part of our lives. **Hebrews 10:25.** We went to church probably six days a week and sometimes if there was something special going on, it was seven days a week. My mother was a preacher, so she was quite influential in the church. She made sure we stayed in church. Of course, in addition to church, the store was part of my life with my grandfather. With my father's work as an electrician, he made money to support the family. However, the store didn't really make money. People bought groceries on credit, and Mother kept a little book with who borrowed and how much they owed. Mother and Father kept me moving with my grandfather. He would run the store in the daytime and then go home to his place and we would run the store in the evening.

Because of the inventory in the store, the store would accumulate a lot of boxes from the canned goods or any other stock that was delivered. My father and mother used to tell me to leave the boxes alone. I was always beating on the boxes like they were drums. I would take two sticks and really beat those boxes like a drum. My parents tried to stop me from beating the boxes, but I just kept going. I believe over time they realized that I had a gift. My mother decided that I needed a drum. So they bought me a drum from the local music store, a mahogany drum. I played this drum all the time. I played this drum until I joined the school band in the fourth or fifth grade.

I then started playing drums in church. I took my drum sticks to church to play. At that time, Shirley Caesar was our choir director in that church. I grew up with her. We didn't know how world renowned she would become. I played drums for her in the church. My Aunt Katie, my mother's sister, moved from Philadelphia to Durham. She had a large number of children, maybe about seven. Her husband had passed away and she had all these kids. He had been in the Navy. They moved to Durham. My mother brought my aunt and her children to our church, and they became a part of the congregation. The choir immediately grew, with my three sisters and my two cousins that lived with us; my aunt and her seven children were included in the choir, also.

Big families are something special. There's never a dull moment and you never have to worry about being lonely. My father and my mother both came from rather large families by today's standards. Because of the time that we lived in, my father's family was all about working instead of much school. My father's mother died when he was young. The emphasis in the household was putting food on the table rather than getting much book knowledge.

Education Not a Priority, Enter Mischief

So I liked school, but I wasn't as good as I would have like to have been. So mischief and getting into a little trouble was soon to follow. When you don't focus on the things that are important, mischief can happen.

Because of my family's store, they had change from the soda pop, as we call it in the south. They would put the money in my parents' bedroom. The change was sorted in little brown bags with quarters and nickels and dimes from the daily sales of the store. My sister was instigating me to go get the money or at least get some money from the bags so that we could go to another store and buy stuff on our way home from school back in 1958. This, of course, was stealing. What a bad habit we had. Normally, my grandfather would take the brown bags and open up the store with that change. So I listened to my sister and started stealing 50 cent pieces or a silver dollar piece from the brown bags. At first the money wasn't noticeable, but eventually I got so much money that I was giving money to my six sisters as well. We had a school bank and we would deposit more than usual in our own school bank account to use for stuff we wanted to buy at school. I remember that my sister and I would bank about 25 cents. That was a lot back in those days; you could get quite a bit of candy with 25 cents. But because we were stealing money, we were depositing much more than nickels and dimes. When the store opened one day, my father and grandfather discovered that they were short on money. They knew about how many coins they had placed in the bags for us to take home the next day. When my teacher called my mother and father to let them know how much was in the school bank accounts, then they realized that I had been stealing money. Where was I getting the money for deposits?

I remember clearly seeing my mother coming down the sidewalk to school that day. She got me out of the classroom; I was in fourth grade. I will never forget it. She whipped me all the way home from our neighborhood school. Then, she called my father and he came home for lunch. When he came for lunch this day, he whipped me again.

I was scared that my grandfather was going to whip me, too, but he didn't. He just talked to me. Believe you me, my backside was sore from my mother and father whipping me. I didn't do that again. I took my frustration and hurt out on my drums. I still played the drums for the church and that seemed to occupy me, but stealing was over with after those whippings.

Sports

I enjoyed sports and, being a boy, I wanted to try my skills and talents and play multiple sports. I remember when I played baseball, and Julius Jones was across the street. He was the coach. I went out for catcher. My grandfather got an old beat-up catcher's mitt and brought it to me. So therefore, I would be the catcher. I went to practice one day, and I remember standing behind the catcher's plate, which is home plate. The pitcher threw the ball and it hit me right in the face. Mind you, this was a hard ball, not a soft ball but a hard ball hit me in the head. Needless to say, it knocked me out and I had a huge knot on my head. I realized, looking back on the incident, it should've killed me at 10 years old, but it didn't. God has His hand on my life even then, to save me from drowning in the swimming hole and then save me from dying from that hard ball to my head. I am thankful to God.

I also used to play basketball. I was not as good as the other guys. I wasn't great, really, but I was all right. The guys would not let me play with them all the time. So what happened was, my father built me a basketball court. He tore down the old garage structure and put up a basket where the garage used to be. That's how I was able to play basketball even better and improve my game. All of my friends who didn't want me to play on the school or other sport team would come in my backyard and play basketball. It also helped me to improve my moving skills especially since I had such big feet, so I needed to make those feet work for me.

There was a tree in the backyard of the house that we now stayed in. I used to climb that tree a lot, to be alone. Sometimes I would climb the tree and just think. There was an old squirrel cage in the backyard. I used to be fascinated by the squirrel cage; it was so old it started falling apart. So the squirrel cage had ball bearings — little small steel balls — on it. One time, I put one of those steel ball bearings in each of my ears. I tried to get them out of my ears by going down in my ear with my finger or something else to poke it out, but I couldn't get them out. I was so scared to tell my mother and father about it, remembering the last whipping that I got. I just left them in my ears for years and years. I was a typical boy, but remember I used to go swimming a lot. So I used to move my head back and forth and I could hear those balls going back and forth in my ear. I was about 11 years old at this time. Pure kid mischief, that's what it was, a boy getting into things.

I also remember telling my sisters about the ball bearings, and I told them to not tell my parents. I thought when I get older and make some money I would get the ball bearings taken out of my ears. I still didn't tell my parents — just walked around with those ball bearings in my ears.

My real saving grace was playing the drums. I've mentioned playing the drums at church, but I also played drums in my elementary school, C. C. Spaulding

Elementary. A friend, William (Billy) Martin, played drums very well. He is deceased now, but we played drums together. We competed in the band for chair positions in the drum section. That's one main reason why I loved going to school, because I loved playing in the band. We played so well in the elementary school band that the elementary band director put us in the junior high school band to play the drums. Oh, wow! When we finally became junior high school students we just blended right in. The band director placed us in the high school band to play drums. We were gifted. We not only played drums in the school band, but we also played drums in the neighborhood. We would look like the Pied Piper marching down the street playing drums. The kids, especially the girls, who became our majorettes, would be following us down the street. We really liked the girls. It was amazing. We did that for years, playing the drums, entertaining the neighborhood and meeting the girls.

Even though I almost drowned, I still loved the water and swimming. I took lessons and became a certified junior lifeguard. Then I kept up with the training and became a senior lifeguard. I never did become an instructor, but I worked as a lifeguard.

One of my best features is I have very big and long feet. Many people, as well as my instructor, would often tease me about my feet and the size that they were. It made me very self-conscious, but I believe it helped me to be a good swimmer. My lifeguard instructor used to look at me with a smile and say, "Boy oh boy, look at those feet." Who knows if it was my feet, the struggle, or the angel who saved me from drowning in the lake?

So between being a lifeguard, swimming, going to church and school, and playing the drums along with other sports, I kept pretty busy. In school, I really found myself in the trade and industrial arts department. My father was an electrician working at North Carolina Central University and my mother worked at Duke University and Lincoln Hospital but was also a nurse, a minister, and an evangelist. She combined all of her skills and gifts into her ministry.

After many years, houses have to be maintained or they will fall apart. Our house was no different, so I was able to help my father in any way possible. He, of course, handled the electrical part as well as some other repairs on our house. Because of my father's work, he was able to meet other men who could also help and he formed a team that included someone to do plumbing and brick laying. So our house was upgraded and remained beautiful.

Being involved in trade and industrial arts in school was a great way to keep me active and to learn a trade. Reading, writing, and math weren't my strong suits in school, but the trade area, carpentry, and working with my hands was perfect for me.

Our entire family were all hard-working people. My mother and father worked extremely hard. They saved up enough money to move us from my aunt's original house in South Durham to another house, across town behind North Carolina Central University, while he and his friends were remodeling the other house. We actually had two houses, and the second house cost around $15,000.

Prior to the move, the store that we owned went out of business, which probably was a good thing because we no longer lived in the neighborhood near the store. I remember in October when I was in the eighth grade, my friends and I were near North Carolina Central University. One day, we went to hunt for pecans. We came across this huge pecan tree behind the North Carolina Central University and they asked me to climb it. I got up in the tree and there was a huge bunch of pecans. I'm up the tree and when I reached for this large group of pecans, the limb I was sitting on gave way and broke.

I fell out of the tree, landing on my back like a pile of bricks. To me, it was like I was floating down out of the tree. How I lived only God knows. After I fell, I must have been lying still and not moving because there was a boy who was with us who went through the neighborhood saying, "Pete is dead. He is dead." My mother had been praying and fasting that day. When my mother rose up out of bed, the little boy came up to the house and said, "Pete is dead." She ran to where I was and she began to pray, "Oh Lord, don't take my son. I've been faithful to you. Don't take my son." That's what she told me. When I woke up, I woke up in my mother's arms. Wow. I said, "Mother, I'm dreaming." She also told me that she had been fasting all day. Just to be on the safe side, the ambulance came and took me to the hospital. My parents and my sister arrived at the hospital, too. What I do remember is in the emergency room at Lincoln Hospital, the nurse was taking x-rays. The doctor said he was amazed that I survived as he was looking at X-rays. My mother and father were there along with my sisters, standing and watching over me. The doctor came and said, "What are those little balls in your head?" Remember earlier I told you about those ball bearings in my ears?

I got extremely quiet because I didn't want to get into trouble. I had never told my parents about those ball bearings in my ears. Well, my sister told my mother and father about the ball bearings from the old squirrel cage. My mother looked at my Father and she said, "What?" My father said, "We're gonna have to get them out." I thought they would have to operate after all and split my head wide open. I was scared to death. Instead, the doctor got a long syringe with warm water in it. He put the syringe in my ear and forced the water into my ears. He suctioned the ball bearings out so far and then took a pair of tweezers and got those steel balls out of my ears and from rolling around in my head. When the balls came out, they were all corroded. It's a wonder that I didn't lose my hearing, but I didn't. The doctor said, "That's amazing!" He also asked me, "When did this happen?"

I told him, "About four years ago."

"These steel balls stayed in your ears all of that time?" the doctor asked.

"Sure did," I said,

I remember one time when a storm was raging: it was a tornado coming to Durham, North Carolina. There was an old garage filled with cardboard boxes behind the house. I had cut one of the cardboard boxes into the shape of wings because I wanted to fly. I have always admired the majestic image of the eagle. Therefore, I perched myself on this barn like a bird and waited for the tornado or storm to help me to fly. Because of my love of eagles, I actually wanted to fly like an eagle, high above all of the cares and problems of this world. My mother and sisters were looking out of the windows, just watching me wait for this storm. My sister reminded me of it later, about how weird that was. Years later I would make an even greater connection with an eagle, but right then, in that season of my life, I just wanted to fly. (*But they that wait upon the Lord shall renew their strength; they shall mount up with wings as eagles; they shall run, and not be weary; and they shall walk, and not faint.* Isaiah 40:31, KJV)

My mother, being a minister, prayed, preached, and conducted all types of revival meetings including ministering to people when they came over to our house. I experienced the power of God through her ministry over the years and I am living off of those prayers even unto this day.

Throughout my high school years, I continued to play in the band, go to church, play the drums for the church and the choir. I was doing things little boys do, but one of my greatest problems was girls. I loved girls. I loved Hillside High School, but after four years, it was time to graduate. What would happen next? Where would I go? What would I do? For me, I really didn't have a mapped-out plan. I believe God has the plan and your job is to obey. I can't say that I listened for His voice as much as I lived my life to the fullest. I'm not perfect by any means but I'm taking every opportunity that is afforded me, good and bad.

CHAPTER 2

UEI and Louisville, Kentucky

I was enrolled in the vocational and industrial arts in high school, learning about electronics. A representative from United Electrical Institute came to my high school in Durham and made a presentation about coming to Louisville for additional schooling. I knew I liked school, but I was honest with myself about my skills, strengths, and weaknesses. Primarily, I was striving to focus on my strengths as a kid and especially as a young man. Honestly, I probably wouldn't do well in college, but I kind of figured that I could go to electrical school. I had taken the subject in high school — why wouldn't it work on the next level? The representative came to the house, sat down with my mother and father, and we heard the man's presentation. They turned to me and said, "Is that what you want to do? We'll do it." He told them the price and at that time it was expensive. So, when I graduated from high school, Mother and Father said, "Well, we'll take you to school." I had never been that far away from home.

We got up one morning, I packed my bags and whatever else I needed, and we started on our journey to Louisville, Kentucky, from Durham. My mother had probably never been out of the state and especially driving toward the mountains. Coming out of Durham, North Carolina, and coming up to West Virginia, there are the Appalachian or Blue Ridge Mountains. I'd never seen mountains like that either. We went through the mountains on the road that would take me to school. It was in the fall of 1969, with the leaves turning to beautiful colors preparing for winter. My mother praised the Lord all the way through those mountains. She saw the boldness, the beauty, the creativity, and the majesty of Almighty God. "Oh my God!" My mother was praising the Lord about the mountains, because nobody could do that but God.

It took us a week for the whole process, including driving to Louisville, Kentucky, getting to the school, registering, and finding housing. After we got there, we were told that there was no place for me to stay in the school itself like a dormitory. You know, I believe that Blacks weren't allowed in any other public housing besides the projects back then.

So we drove around Louisville and start to ask Black people about housing. That's how I ended up at YMCA.

Some local people told us that's where I could stay. They had boarding rooms at YMCA on 10th and Chestnut. I had a room on the second floor. That's where I stayed for $10 a week. I had a roommate and his last name was Johnson. He was in mortuary school.

Since my father was an electrician, I should have been a natural. I did okay in high school, but one reason it didn't work out that well for me was that I was color blind. Being color blind made it very dangerous for me and the other people I worked with. So much of that work is based on how things look and by color. If I couldn't tell the correct color of things, it could be a disaster so I just didn't do well and make good grades in that school.

So I stayed in Louisville from 1969–1971. Unbeknownst to me then, the public library across the street from the YMCA had a training program for Black women, as librarians. They were being trained and certified to be librarians for the entire south at that location. Who knew? I didn't know it at the time but there were a whole lot of women there, and my weakness was women.

So I could stay in the YMCA but I needed a job to keep paying the weekly $10 fee. I got a job at the Commonwealth Insurance Company as a janitor. I also was the first African American or Black, at that time, to work at a place called the Normandy Inn, down at the riverfront of Louisville, Kentucky. Louisville sits on the

Ohio River. There were these White guys that worked at the restaurant; I hadn't interacted with White folks before until then. These guys got me the job as a salad boy. I just needed a job, but it was big news in Louisville for me to work there. The Black-owned newspaper, *The Louisville Defender*, wrote an article about my new job. I was the dishwasher at first. The other young men I worked with saw the leftover food that was coming back. So we ate the good food that customers were throwing away. Normandy Inn served food like steak and lobster. The salad boy would get me out of the kitchen, so I wanted that job. They taught me how to make salads and thankfully, I got the job. I didn't tell my parents that I wasn't doing well in school. I hid it from them and kept on working. Louisville was the place I spread my wings. As I stated earlier, I spent so much time in church that I wanted a taste of the world and fun. **Galatians 1:6.**

Phillip Morris — Laborer

I hung out around Shawnee Park, at the skating rink called "Fountain Ferry." I went to these events to escape. We would meet the young women from Central High School next door to the YMCA. I had to have money to be with the girls. I was really what you would call wild. I would spend the night with women and in the morning have to climb out of back windows. I was forever getting into some type of trouble with or over a girl. I was just out there. Because I had a little money, I would travel to Paris, Kentucky, which is near Lexington, meeting friends and other young women there. I was a janitor, working at Commonwealth Insurance in the evenings. Next, a young man asked me if I wanted a job at Phillip Morris. Tobacco was king in Kentucky; Phillip Morris was a huge tobacco packing company.

I said, "Wow, and yeah." It was more money than I was making as a janitor, so I went in for an interview and got the job at Philip Morris as a laborer. That was very good money back then. Mind you, my mother and father didn't know I was working. I was doing what I wanted with the money because Mother and Father were still paying my way to school. I got tired of lying to them. I got tired of taking their money. I decided to get a job so I could make my own money and do what I wanted to do.

The Green Volkswagen

I went home to Durham for a visit. My father helped me get a green Volkswagen in 1969. I had girlfriends everywhere I went, including North Carolina. While I was in Louisville, my girlfriend in North Carolina was graduating from high school. She was going to school in Nashville, so I drove to Nashville.

I thought we'd be able to really get it on while I was there. I had plans for a lot of huggy huggy and kissy kissy, especially since she was far away from home. We were planning to do big things while I was there. In the old days, they called it "making that move." That trip we were "gonna make that move." I took my little Volkswagen to meet up with her and her girlfriends. I remember the girlfriends out on the porch just waving and smiling, so happy to see me. We were going to go check into a motel to get it on. But that never happened. I told her that was all right and took her back to school. When I went back to Louisville, I started my running women as usual. I kept working at Phillip Morris, making money so that I could keep up my lifestyle of partying and having fun.

My fun was about to come to a screeching halt. One day I came home to the YMCA, still in Louisville, and Mr. Jamison was the guy at the front desk. I got a letter, and I didn't open it up at first. After I got settled and got near the phone, I called my father and he said, "Hey, Pete, what are you doing?"

I said, "Uh, just working, just working."

He said, "I got a letter here from your Uncle."

I said, "Which uncle?" I started naming all of my uncles and Father would say, "No," after each one that I named until he finally said, "It's from your Uncle Sam. You've been drafted into the Army."

I said, "I don't want to go to the Army!"

Back then I definitely didn't want to go to the Army because of the Vietnam Conflict. Plus, some my friends were getting killed in Vietnam.

CHAPTER 3

Drafted into the Military

I continued to say to myself, "I can't go to the Army now." I was determined to get out of going to the Army, so I went to the Federal Building to talk to the recruiter. His office was located on Broadway in Louisville. He basically ignored my request to get out of going and instead said, "We have one slot for the Air Force, and you and someone else are eligible to take the test." I didn't want to go to the Army so I had to pass the test for the Air Force; otherwise, I would have been in the Army. I took the test and the recruiter said, "I'll let you know this evening." That evening, lo and behold, I passed the test — with a high score, too. I guess it was decided for me that I would join the Air Force because that was the one slot open to me. That was a move and favor of God right there in itself. I got drafted and I took the physical. I was still trying to get out of going because I knew I had falling arches in my feet. I thought that the falling arches would get me out of going. The recruiter said, "Oh, we got boots. We can build those arches up. Your arches won't be falling after you get the boots. There's no reason why you can't go." I was disappointed again that I couldn't get out of it, but I had no choice. I had to report that next day, probably Monday or Tuesday. So I said to myself, "Well, what am I gonna do?" There was nothing left to do but pack up my things, check out of the YMCA, drive myself in my Volkswagen back to Durham, North Carolina. From there I said goodbye to my parents and got on a plane. It was Piedmont Airlines back then. It was the first time I had flown on a plane. I left Durham to get back to Louisville to get sworn in by the ecruiter. After getting sworn in, I got on another plane.

San Antonio, Texas

When the plane landed, I was in San Antonio, Texas. A bus took me to Lackland Air Force Base for basic training. So I did my basic training in hot, humid San Antonio, Texas, at Lackland, AFB. As training went on, I became sharp and had an outgoing personality, so I knew everybody in the barracks in a short period of time. In the Air Force, there were squad leaders and I decided that since I was here, I wanted to be a leader. I knew I didn't want to be the grunt where somebody was always telling me what to do. I wanted to tell others what to do. The training instructor, better known as the TI, was watching me very closely. I realized that he was watching me so when we were marching, I stood with my chest out, marched really strong and sharp. So the TI said, "Buie, come on up front." I guess it was so that he could watch me even closer. Over time marching up front and seeing how I handled myself, he made me a squad leader. I really liked that. To make a long story short, we finished basic training and I graduated. I went from Lackland Air Force Base to Chanute Air Force Base in Illinois for technical training. I was admitted to a training program to be an Air Crew Life Support Specialist. An Air Crew Life Support Specialist inspects all of the life support gear for the pilot. I got that training, which was an ordeal and a journey all at the same time. My high school training, my father, and even the work that I did in Louisville were skills that added to my military journey. I now know that it does all work together for a purpose. I didn't think so then, but it did work for me. **Romans 8:25** . One of the things that truly kept my spirits up through this whole journey was listening to music that I never listened to before. At home, I listened mostly to church music. Now, I could listen to whatever I wanted to listen to and I enjoyed it. After Chanute Technical School in Illinois, I was an Air Crew Life Support Specialist, and next I was sent to Hill Air Force Base in Utah.

Utah

When I got to Utah, compared to Durham, it felt like being on the other side of the world. I was doing my job, inspecting the life support equipment, and I was enjoying life while running the streets of Ogden, Utah. My enjoyment and weakness were girls, girls, girls. My buddies and I would meet girls from Ogden to Salt Lake City. That was always my problem. We had a Recreation Center, that's where we'd meet and drink beer while the jukebox was playing. I'd get behind the jukebox and turn it up and down. Even then, I was acting as the disc jockey in the Recreation Center and they always said that I was pretty good at that.

I enjoyed several months of partying and meeting friends in the Rec Center and acting as the DJ. Now it was time for another assignment. I was being prepared to support Vietnam activities but also going to be stationed in Thailand. We had to go to the Philippines, for Jungle Survival School. It's amazing that just how the military prepares for you for your next assignment, God does too. Just the sound of Jungle Survival School kind of scared me, but I was in the Air Force, and my "Uncle Sam" told me what to do. **Exodus 3:10.** I was given an opportunity to go home to Durham prior to shipping out, so I went to church while I was at home. While at church, they prayed for me, for God to protect me and keep me safe. There was this older lady, Mrs. Plumber, who lived up the street from my parents' house on the corner. I told her that I was going to Thailand, but first I had to go to the Philippines for training. She gave me a little Bible. She said, "Whenever you need God, hold this Bible." It was a small New Testament Bible. She said, "Just pray over it and God will keep and protect you. We will look forward to all of the testimonies that you will tell us just based on this small Bible." The Air Force issued us all duffel bags, and so I put the Bible in my duffel bag. I was covered by prayers of loved ones. Off to Jungle Survival School I went.

CHAPTER 4

Duty Station, the Philippines

I arrived in Jungle Survival School in the mountains in the Philippines. We were up in the area where the native Filipino people are called "Negritos." I don't how they did it, but they stole all my underwear. I didn't have any underwear. Anyway, while I was in jungle survival school, we literally learned how to survive in the jungle. We practiced dodging bullets and bombs, crawling under barbed wire, and hiding from the Negrito instructors — who would eventually be the enemy in war time — and doing anything else that would help us to survive. While we were in school, there was a young man who took a liking to me. He used to sleep under my hammock in the jungle and protected me the whole time I was in jungle school. We used to be trained by being able to hide in the jungle. The natives would be rewarded with a bag of rice if they could find us. That was part of the jungle training. We would crawl under the underbrush and they would simulate an actual attack with bombs, etc., going off because this was Jungle Survival School and we're trying to survive. I did that.

Thailand

My next duty station was in Nakhon Phanom, Thailand. When I got there, I was in the 23rd Tactical Fighter Squadron. My job in Thailand was a combination of the Air Crew Life Support Specialist as well as a Jungle Survival Instructor. We inspected all the equipment that could have been used in the Air Force and assisted training others with their jungle survival skills. This is also when I was introduced to marijuana. I had never smoked marijuana before until I got to Thailand. They called it "Buddha" over there. I smoked my first hit of Buddha and I thought I was going to die. I started hallucinating. It was so powerful. I remember crawling up the hill in the jungle. I grabbed my little New Testament Bible and started praying. I held this Bible real tight and prayed, "God, take me through this." I prayed and that feeling just went away.

I had a girlfriend down on the economy, which means off the base. Her name was Pawn, and she had a little boy. By now, as a Jungle Survival Instructor I was teaching others how to survive. I was teaching them how to handle multiple jungle encounters, especially after they were ejected out of an aircraft and shot down. My girlfriend Pawn really liked me, and I enjoyed my time while in Thailand. They called me "Buck Pete." I bought a lot of clothes because clothes were cheap there. I was still smoking Buddha or a lot of marijuana in that time. Back then they subjected us to a urinalysis test. I was smoking so much I knew I wouldn't pass the test and was trying to dodge it. I knew I couldn't dodge it forever. So, I had my roommate pee in a bottle so I could pass a urinalysis test. I kept smoking weed and then started smoking heroin too. I was in bad shape, really strung out on drugs. My roommate continued to pee in the bottle so that I could pass the urinalysis tests for months. I remember when it was time for me to leave Thailand, I thought to myself, "I would love to take some of this heroin back with me to the United States." You know that I was addicted if I was willing to risk everything and bring drugs back with me. So Pawn helped me stuff the Buddha in cigarettes. I smoked Kool cigarettes. So having worked at Phillip Morris, I knew how to get the tobacco out of the cigarette. I replaced the tobacco with Buddha and heroin. Then I put the cigarettes back into the pack just like normal.

We used to put the cigarettes in these little plastic containers so they would stay fresher longer. I had come out of the jungle and was staying downtown until time to leave. The Freedom Bird, a large 747 aircraft, would take me back to the United States and the real world. At the airport, I had to go through customs. Mind you, I had that heroin in my cigarette pack. When I was going through customs, the guy looked at me and I was so nervous and looking around suspiciously. I guess I was very obvious. They went through my things and took the plastic container that held the cigarettes and pulled the top of it back to look inside. My girlfriend, Pawn, had trimmed the cigarettes and were the exact same level as the other cigarettes so they looked the same. They didn't have drug sniffing dogs to sniff out the drugs from the tobacco; they just went on sight. So I was able to get on the Freedom Bird, come back to the U. S., and landed at Travis Air Force Base in California. I looked so bad, and there were anti-war demonstrations going on with young people protesting at the airport. People didn't pay any respect to military soldiers back then as they do now. We weren't welcomed back in our own country.

When we came back from Vietnam, my friends and I were kind of like junkies from all of the freely available drugs, as well as the other things we were exposed to. When we got to the next base, I got that heroin and my so-called friends and I smoked it all up. I looked in the mirror at myself and saw myself for who I really was and that I was actually addicted to drugs. I knew I wasn't raised that way, and I surely couldn't go home the way. **Romans 7:13-24.**

I was reassigned to Hill Air Force Base in Utah. When I got back to Hill Air Force Base, I checked myself into rehab so that I could get myself together. That one-year tour of duty in Vietnam from 1972–1973 had an impact on me that I will never, ever forget.

CHAPTER 5

On Leave to North Carolina

God helped me again. **Romans 8:28.** I was able to get clean before I went back home to North Carolina. I arrived home and saw all my family members, who were so glad to see me. I went to see my girlfriend, and she literally broke my heart. She had first told me that she would wait for me while I was in the Air Force, but she decided that she couldn't wait on me and she was in a relationship with another man.

The Birth of the DJ, AWB

After my leave, I returned back to my duty station at Hill Air Force Base in Utah.

Even though I was clean from heroin, I wasn't clean from the partying lifestyle. I reconnected back with the women, clubs, and partying in Utah. In my eyes, I was having a good time. Then they had a dance at the golf course on base. I went to the dance and I acted as a DJ in the Rec Center. They used to tease me anyway and call me the DJ. It seemed like I had a natural gift to be a DJ. The club was called the Brotherhood. Their regular DJ had to go on emergency leave for something and they didn't have a DJ at the NCO (Non-commissioned Officers) Club. They asked me if I wanted to sit in, because they heard about me working the jukebox years ago. I said, "Yeah, I'll do it." So I did the dances along with my roommate. My roommate stayed downtown in Ogden, Utah. We had an apartment, and we would go back and forth off post to on post for these parties. It was now 1974, and my roommate said that he met these girls and we were supposed to go to their house to meet up with them. He said, "I want to introduce you to this girl. I really want her for myself but I'll let you have her." So we went over to the girl's house. We started playing Bid Whiz and I pulled out a joint to smoke. I wasn't on heroin but I still smoked weed. This girl named Jeannette said, "Not in my house will you smoke weed." So we went outside to smoke. Mind you, I was still in the military at the time but still doing my thing. When we came back in the house, Jeanette had these two little kids, James and Trina. I think she was really trying to run me away because she had these two kids. I knew she had kids because she brought the kids in to meet me. I saw something in these two little kids and I saw something in her, too. Jeanette was different. So,

I started dating her and not too long after we started dating, we got married. It was a small wedding at her mother's house, but I was married just the same. I met her mother and the whole family. As a matter of fact, I was the lifeguard in the community in Ogden, Utah. So I knew a lot of people and I was rising disc jockey since my second time in Ogden.

I was getting ready to make my first public appearance, not at the Brotherhood Club this time but the NCO club. I realized that I needed a name for myself, an entertainer name. I had signed a contract and I said, "I need a name." This girl came in the shop and we chatted. She said, "Your name should be AWB."

I said, "I can't be AWB because that's the Average White Band. And she said, "Yes, but make it AWB but add the "Soul Express.'" So that day I became, "AWB Soul Express." So on the main billboard sign was "AWB Soul Express." When everybody saw that, they thought the Average White Band was actually coming to the NCO Club. They didn't get the part about the Soul Express, but it worked! There was a line out the door and down the street waiting to see the Average White Band because it was free. Unbeknownst to everybody, this Black man shows up on the stage with two speakers and a turntable and not a full band. What did I do? I rocked the place. I rocked the place out. That's how I became AWB. That's where I was named AWB with the tag "Action. Wonderful. Boogie." People all over from the local area would come to hear me and my music at the NCO Club.

I got a job at the American Legion in Ogden as a DJ. I then got a job downtown in Salt Lake City playing at the Elks Club. I started playing all over the place. I had so many jobs, I needed bigger speakers and a faster, better turntable.

My wife at the time co-signed for me to get a loan. I bought some big speakers and some turntables, and I was able to fulfill all my contracts at all the club parties they wanted me to come to. I was a big name. I had AWB Soul Express, played the top 20 hits every week in the community so that people knew the number one hits. There wasn't a lot of Black music or entertainment in Utah. I was playing everywhere. While interviewing on TV and on radio, they would ask, "Who's AWB Soul Express?" I still had my job in the military in addition to the DJ gigs. Being a DJ was only a part-time gig. My boss told me, "If you take off 'Soul Express,' you can probably get into the doors of White establishments." After that, I became AWB Disco Magic instead of AWB Soul Express. I was then able to go into the Officers' Club. I was able to appeal to an even wider and more diverse audience, and other parties. I made a lot of money and just played everywhere. I was married and still AWB. I don't know what I was thinking, but I still had girlfriends and even invited them to my wedding. Remember, my wife Jeanette had two kids, a little boy and a girl, James and Trina. Jeannette had a cousin named Elaine, and she had a daughter named Leslie. Leslie used to babysit for us. Elaine became sick and she asked that if anything happened to her, would we raise Leslie as our daughter. We agreed to take care of Leslie along with the other two kids. After Elaine died, we became foster parents of Leslie first.

Then after that, it was time to be reassigned. I had a new assignment to Germany and that meant we couldn't take Leslie with us while she was in foster care. We didn't want her to go into the system, so I adopted Leslie, Trina, and James and gave them all my name. So I was now married and immediately had three children, at age 24. Trina was 18 months when I met Jeanette. James was about three years old. Leslie was 11 years old and she was half Italian and half Black. During that time, Jeanette got pregnant, and we had a baby together. Her name was Jena. Jena was actually a miracle baby, born premature. We had everyone pray that she would be healthy. Of course, the Saints prayed and Jena lived and was our living miracle.

No time to settle in and get even more acquainted because it was time for us to go; next stop Germany.

CHAPTER 6

Duty Station, Germany

In 1977, I was married, with four kids, and in the military. Our new assignment was in Germany. What an adventure? I was still a part of Air Crew Life Support as a Specialist. I had to go by myself to set up house in Germany before the rest of the family came over. I would have time to explore and live in this new country before they came. I was in Germany for the military but I was working in the evenings at the NCO Club even in Germany. AWB was now international. I was playing everywhere all over Germany. I went to Luxembourg; I played everywhere and even got a contract at a club called "Big Bird," as well as so many clubs I can't even remember them all. I really had two full-time jobs, one in the military and the other as an entertainer at night. Wow. I really played everywhere.

We were actually in Germany for three years. I travelled all over Europe, because it is very easy to travel there. I went to Italy and Pakistan and enjoyed all of the sites. I was very surprised at the level of poverty in Pakistan. It was amazing how people live in other countries. Things we take for granted here in the U.S. are truly a luxury in other places.

My Mother's Illness

My mother got sick while I was in Germany in 1977. She had a brain tumor, and they didn't think she was going to make it. So I had to go home. I went home, and saw my mother, and checked on my father and everybody back in Durham. I eventually came back to Germany and, miracle upon miracles, God restored my mother's health. He restored her from a brain tumor, imagine that. The doctors were able to take the tumor out and then she was restored. Until one particular time she was preaching, and I was told she fell out. Boom. She went back in the hospital and I had to come back home to Durham again, this time with the entire family. Before coming back home we had to pack up and leave everything, including DJ equipment. My friend Charles Weaver had power of attorney for everything, while still in Germany. My mother was in the hospital and it didn't look good this time. I'll never forget it. My father and I went to the hospital to see her. She was holding on to see me before she passed.

There were deep rooted thoughts going on in my mother's heart. She was hurt. She was crying. She had written down the names of people that had hurt her in the church. My mother was a State Mother and powerful woman of God. She did revivals and ministered mightily. She passed away and I remember holding her hand. I said, "Mother, you ready to go?" Because she was suffering, she looked at me, held onto my hand, a tear came down her eye, and she left. She went on home to be with her Lord. My father was in the room along with us to say goodbye.

After my mother passed, we went through all of the procedures for the funeral and had one of the biggest homegoing services that I have ever seen in Durham, North Carolina. Shirley Caesar sang at my mother's funeral and blessed us all. What a memorable homegoing celebration. **2 Corinthians 5:8**

During this same time my sister, Arhonda, became distraught over my mother's death. She was missing for at least seven years. Ultimately, they found her remains, and she was finally laid to rest. Grief is horrible and can at times overwhelm you without the Lord and possibly some professional counseling.

Just as there were sad times and horrendous grief, there were also some good things happening as well. My youngest sister, Donna Buie, auditioned for the movie *The Color Purple*. Donna got a role in the movie along with Whoopie Goldberg and Danny Glover. She played the role of the young woman that married Whoopie

Goldberg's step-father. Our family was super proud of her accomplishments.

I grieved the loss of our mother, but as a soldier, there was only so much time I would have with my family. I knew that there would be another assignment. I was kind of held up and in limbo because of everything that happened with me and my family leaving Germany, etc. I contacted my commanding officer and after some negotiation and discussion, my next assignment was Barksdale Air Force Base in Louisiana.

CHAPTER 7

Duty Station, Louisiana

I was stationed at Barksdale AFB, Louisiana in 1981, for the next three years. As usual, I set up housing and looked over the landscape of the base while my family stayed in North Carolina. After things were in order and I got things ready, they would move to Shreveport, Louisiana. I was still working in air crew life support as a specialist, inspecting survival equipment. There were a lot of drug problems in the air crew life support area that I was not involved in, thankfully. My commander and the first sergeant knew me and knew what kind of person that I was and asked me if I wanted to run the dormitory. I said, "Sure." So they took me out of Air Crew Life Support and they busted those people in the area doing wrong. I was out of there and away from that drama. I now had the job of running the dormitory or housing for soldiers. As usual, I got a job at the NCO club as AWB. I was still the disc jockey. I would get to do all of the parties, dances, and special events.

One day, one of my friends, Rick Anthony, said, "Man, they are hiring at the radio station."

I said, "What?" He knew I had the skill because I had done it before. I just needed to get in the door, and my boss made sure of that.

Radio

Everybody always loved my voice. So we went to WDKS, known as Kiss radio 98.1 FM. I had an interview at the radio station with Steve Scott; the Rock Star is what he called himself. All I said was, "Mr. Scott."

He said, "You're hired." I looked at him with the strangest look on my face but realized he was serious and he gave me a job as a radio announcer. He liked my voice, just like my boss.

In addition to the job, the station manager gave me a new name that was different from my AWB DJ name. I was named "the Sarge on the Radio," and I was really good. He wanted me to be quiet, just to give the ID of the station at the top and bottom of the hour, announcing the news and any public service announcements. That was all they wanted for me to say on the radio and not be so loud as the AWB Disco DJ. I believed that they were trying to hold me back. But that was the weekend jock's role. I played at midnight to six in the morning and I had a great time. I became very popular. I was at the radio station in Louisiana from 1980–1983.

The military put my house on the main entrance and I had another job giving tours or directions because of my access at the entrance on Barksdale Air Force Base. I had that big house on the main drag. I was blessed by that. I give God all glory and honor.

In spite of all of God's blessings, I still hadn't let go of the different women in my life other than my wife. I am not trying to offend anyone but want to be honest and tell the truth. There may be a man who is struggling with this same problem I experienced. I pray that my mistakes and testimony will help someone be delivered.

I kept my secret life hidden. I didn't know if my wife actually knew it or not. Amazingly enough, no women came to the clubs or places where I worked and caused fights until one night.

A lady got jealous and she reported me to my base commander. She told the commander I was having a relationship with her and that I was married. She even claimed that I gave her a sexually transmitted disease. Back in those days, if you went to the Health Department, they would ask who you had been in a relationship with, so this time, I had tell my wife. To save her embarrassment, I took her to a clinic that was off base, so that she could get examined and then get a shot. I had to tell all these other ladies the same thing.

My wife didn't know about all of the others, only the one who threatened to tell her and expose me to everyone.

I didn't deserve it, but my wife forgave me. This is the first time I've ever mentioned this to anybody in detail besides my wife. She knew about it, and she forgave me. After that incident, you would think I wouldn't do anything with anybody because I had a good wife, Jeanette. She was a loving and a giving person. When I would get into trouble, she would help me get out of all of it. Even after she found out the things that I did with the one woman, she still forgave me. It is no excuse, but I was in my early thirties and when you're young, you're not thinking about the ramifications and all of that. You are just thinking about what your flesh wants at the time.

But here I go again. I had another young girlfriend who my Black commander wanted as well. That was a mess and somehow I was still about to be employed at the NCO Club, acting a fool and acting crazy in spite of what I had just gone through. As time went on, somehow I was reassigned out of Barksdale, Louisiana, to Offutt, Nebraska. My commander/supervisor didn't know how I got assigned to Nebraska. Only God knows how, but clearly, Nebraska here we come.

CHAPTER 8

Duty Station, Nebraska

Nebraska is an agricultural state filled with wheat and land and wide open spaces. As usual, we packed up from Louisiana; Jeanette, three kids and me going to Nebraska. Leslie stayed in Louisiana, because she was now married and had started a life of her own. On our way, we decided to have some fun at Six Flags and enjoy the other sites in Texas. A funny thing happened on our trip while we were transitioning out of Shreveport, Louisiana, Barksdale AFB, Louisiana, and then on to Nebraska. We needed some clothes, and had a lot of money to spend. We went into a Playboy Club clothing store. I didn't know or understand anything about Playboy. Just out of necessity, we got the kids dressed up in Playboy gear with all of these bunny rabbits on their clothing. It was hilarious until somebody said, "Those are Playboy clothes!" So we got a big laugh at that.

During our trip to Nebraska, we saw accidents while driving that far and saw actual deaths on the highway as we drove. It was a teachable moment for the kids, because we got to tell them how God protected us. That's the good part about the travel.

We finally got to Offutt, Nebraska. I was working in the Headquarters SAC, or Strategic Air Command, which is one of the most dangerous areas in the military. I was in charge of the front desk operations (hotel management), and I had to be on call all 24/7. I was a team leader and got a TDY (temporary duty assignment) to Korea, field operations. I was in charge of the laundry tent. So I ran the laundry and made sure all of the soldiers' clothes were clean and then saw that the right clothes got back to the right person. While running the laundry tent, one of the team members was attracted to me, and I was attracted to her. We got caught up in a relationship and had some drinks, got drunk and, we fell asleep in the laundry tent. I woke up and there was the police and I believe the first sergeant.

Of course they reported me to the commander, that I was in a cohabitation relationship with a female and in the wrong place. I told a lie that I fell asleep but I didn't cohabit with her, but I did. There was a commander there who wrote a letter on my behalf to protect me. I was a married man, and thus my whole military career was on the line. This was also something else that I kept away from my wife.

Oh, the favor of God for this commander to protect me in spite of what I was doing. **Matthew 28:19-20.** They moved me out of hotel management and moved me into the dormitory. I was running the dormitory there in Offutt, Nebraska. I did very well in that job. I had a chance to be interviewed by generals and officials associated with SAC. They were very impressed by the quality of life for troops living in their dormitories.

I was still partying and being the DJ AWB. I had a relationship with one of my coworkers, from the previous time when I lived in the area. When I came back to Offutt NE, she said that she wanted me to see something. She had a boyfriend, so I didn't suspect anything of it. I went over to her house and she brought out two little twin boys for me to see. We had a relationship as well, so I really didn't know if those boys were mine or not. Even though she had had relationships with other men, she insisted that these boys were mine. I kept that quiet the whole time I was there. I knew that I had relations with her but the children being mine, I didn't know. It always bothered me about those two boys, but I just went on with my life as it was.

A colonel asked me once, "Where do you want to go next with your career?"

I said, "I wanted to go to Alaska." My colonel got me assigned to Alaska. So prior to going to Alaska, the entire family and I took the scenic route from Omaha,

Nebraska, even to Washington, DC. We had a great time and experienced all there was to see. Next we drove down to North Carolina and Georgia and had a good time visiting family. Then we spent time at Myrtle Beach, South Carolina and came across back to Louisiana. We visited my oldest daughter, Leslie, because she was pregnant. Then we came across the country headed toward Alaska, while partying and having a good time. But first we made a stop in Utah because that's where all of the kids were born. We went to the amusement parks and told everybody to come on, let's go, because I had money as I was traveling here and there on Uncle Sam's dime.

My wife, my kids and I had such a good time on vacation. The money was getting low, so it was time for us to go. I had about $350 and we still had to get to Alaska. We didn't have enough money to really stay at hotels on the way so we primarily slept in the car most times. My wife said, "Well, you drive; I drive at night and you drive in the daytime." So we drove all the way up through Montana, across the United States, across the Canadian border into British Columbia, and into Alaska. I would sleep while she drove and vice versa. I remember one particular day; I went to sleep and she drove. As you get closer to Alaska in the months of June and July, you have more daylight. Alaska has almost 24 hours of daylight in the summer. The winters have the long nights. So my wife was driving all day, and she didn't look up at the signs and I woke up and we ended up in an Indian village. I looked around and said, "Where are we? We're 200 miles off course." Remember we were using old fashioned maps to travel. In this village, people had never really seen Black folk before. The people came and embraced us and made us feel good. But we had to get out of there and get back on the right road and going the other way. There was no GPS back then, just maps. I remember we had crossed over into Canada and saw the Great Slave Lake. Quite a majestic and beautiful scene was Slave Lake. We crossed over the border from Canada back to the U. S. Mind you, during that time, money was low, so our kids were using the bathroom on the side of the road, because we didn't have money to stay in a hotel. So we drove literally three and half days to get to Alaska and Eielson Air Force Base outside of North Pole, AK. It was amazing to see all of the wooded area, to see how majestic the mountains were. We saw the steep drop off cliffs. Looking over the cliffs, we saw horses running wild and in other parts there were buffalo, tall trees and so much greenery. Alaska still is the wild and open frontier of the U.S.

CHAPTER 9

Duty Station, Alaska

I finally checked into the base. We stayed downtown until we got base housing after spending a short time in temporary housing. My skills in the dormitory continued to follow me to Alaska because the commander heard that I was good at running the dormitory and taking care of the troops. They asked me what I wanted to do, work in billeting (hotel management) or run the dormitory. I said, "Run the dormitory." So I got a job running the dormitory and I was still able to be a DJ again as I had all over the country and abroad.

At this duty station, I was getting letters, emails, and phone calls from the same lady with the two boys. It had never been determined whether they were actually mine. She was threatening to blackmail me and tell my wife. I was trying to keep that all under wraps and dodging and hiding certain things to try to stay calm. I told her, "I don't want you to tell my wife." I knew deep down that if this woman told my wife about her and these boys, my marriage would have been over. One particular night, I was actually out with my wife and got tipsy. Keeping the secret was overwhelming, to say the least. My deeds were going to finally catch up with me and possibly for good. **Numbers 32:23.** So, my wife and I were at a bar and my wife told me, she said, "I just love you. Regardless of anything that you've done, I still love you." This was not going to make it any easier, telling her about this lady and her two boys. She trusted me and had forgiven me for the last time and now this. But I had to come clean. So, I looked at her and I told her about the boys. Her mouth just fell open because I don't think she was ready for that. What woman is ready to hear that? It took her by surprise. I took a great toll on her over time and literally a lot of the life and energy she had in her. I saw her lose a lot of weight. She stayed up in her room and she told me something that I never really could say until now but, "That woman robbed my cradle. She robbed my cradle." We went through her pain together. How we went through that and actually stayed married, I will never know. Over time, we got better and we got stronger.

The family lived on base. Then my daughter moved from Louisiana to Alaska because she got divorced from her husband. She moved into on-base housing. In the meantime, I was still going back and forth with this woman over these boys and still trying to keep her quiet. The lady with the two boys was planning to marry someone else. This went on for years and she finally went on with her life. Buying this house for the family meant that I was still playing in the clubs for the extra money to be able to buy a house, in North Pole, Alaska. One particular night they had a battle of the DJs.

A guy was leaving to go to a new duty assignment. He had the number one slot at the club. I wanted to be the DJ and get in the #1 slot. We battled for it, and I won the position. So I was a disc jockey for years all around Fairbanks and anywhere that wanted me in Alaska. If it was ladies night, it was a little dangerous for me, given what I'd been through and with my usual drinking and partying, but I was still in the military. God still protected me in spite of myself.

But while in Alaska, I was still in the Air Force, a disc jockey, playing all over the state, and even played in country bars. I was "Anthony Western Boogie." I had a real following wherever I went. People used to laugh at me because I took on the whole cowboy routine, hat, cowboy boots, belt, etc. I built this whole entertainer personality, the name and the whole nine yards.

They laughed at me, they kept on laughing, but I was laughing all the way to the bank. Each time and place that I played, I was amazed at the response of people to the music I played. I didn't mention it earlier, but I became a Mason while in Germany. I was a member of the Masonic family, and I came here even more active in Masonic community while in Alaska. At times I recruited for the Masons. I was bringing in my friends, and specifically my three best friends, to become Masons. I was still AWB the DJ, having what I thought was fun and being crazy at these parties.

Once the Masons sponsored a gospel concert and the special guest was Pastor Shirley Caesar, from Durham, North Carolina. As I mentioned, Shirley Caesar was our choir director, at my hometown church that I grew up in. She sang at my mother's funeral. I also played the drums for her years ago, at the church. So, we were able to bring her all the way to Alaska for a full line concert as a fundraiser. We all enjoyed it very much.

My wife was going to church in North Pole, Alaska and taking the kids to church. My grandson had special needs and was in a wheelchair. We had to pick my daughter and her son up to go to church. When we got there, we would literally have to pick him and the wheelchair up to get him inside the building. There was no handicapped ramp so it was steep incline. I remember going to church and sitting in the back seats, just passed out asleep from the night before. There were many, many days that I was home and my wife had to go to church by herself. She went to church alone in Germany and everywhere, especially because I would be home in bed. Sometimes I would go to church in the other places we were stationed, but in Alaska, especially, I was in the back of the church asleep. You know I think about it now, and compared to my life back then, it was just sad.

I kept playing, doing gigs as a DJ and kept those other jobs so that I could prepare to retire in 1992. I was in Alaska when I actually retired. I had a wonderful retirement celebration, and the very first thing that I did after the ceremony and when I was released from the military was to roll up a big joint and smoke it.

I didn't stop; I continued smoking weed, acting crazy, going up to Native villages, playing for schools all over Alaska, and working at the radio station. I was an account executive in sales at a radio station in Alaska. They had a big promotion and they needed a motorhome. So as a DJ and knowing a lot of people, I contacted a car dealership. They let us use the motorhome for the promotion that the radio station was having. Because of my connections, communication skills, and enthusiasm, they offered me a job at the radio station. I really struggled as an account executive, because reading and writing was very challenging from an early age. Thankfully, my wife, Jeanette, said she would help me do a good job and she helped me write the commercials. I was good at talking and people loved my voice. I knew how to market the radio station.

I was still deejaying, playing music, and definitely getting drunk. Sometimes I would come back to the radio station smelling like alcohol, and sometimes even passing out. I didn't know my boss at the radio station was the owner of the station. He told me, "If you keep on going the way you are going, I'm going to piss on your grave." Those were his exact words. I was smoking cigarettes and drinking, and living an unhealthy lifestyle. On the outside it appeared I had stable life, owned a house, had a loving wife, kids, and a little money. It looked like a good life to the outsider. But everything else was in chaos. I was angry, started arguing with people, and doing drugs. It was not a good life and I was putting my job at risk.

One particular time, I had to fly to the Tanana Native village, in the mountains of Alaska. I played there, stayed overnight, and had to fly back in the morning for a radio promotion. I was trying to get back on the plane with all of the equipment. I couldn't get my equipment back because of the excess cargo already on the airplane. My good friend, Darnell, was with me, as a roadie. He said, "I'll stay back." So he stayed with the equipment and loaded it up on the next flight out. Thanks to Darnell, I was able to make it back so I could be at the radio event that I had lined up.

I had a radio promotion downtown. Normally I would go by myself to these events, but my wife,

Jeanette, was waiting for me and she asked, "Honey, can I go with you?"

I said, "Yes." So she and I got in the car and went down to the radio promotion at the Greyhound Lounge in Fairbanks, Alaska. There were some people there who had never met my wife. She had always stayed at home raising the kids, so people enjoyed meeting her. Jeanette enjoyed the excitement of the event and seeing the people win prizes.

Even though I was enjoying myself with my wife and family, those boys being with the other woman and her trying to blackmail me, was still in the back of my mind. I promised that if those boys were mine, I would want to meet them in the future.

So after the radio promotion event with my wife, we grabbed a hamburger at the C & J Drive-In. We were feeling really good and hoping to enjoy each other when we got home. We were feeling good. While she was in the bathroom changing her clothes, I heard Jeanette scream. I ran in the bathroom, and she had gotten sick and thrown up all over the walls and everywhere in the backroom. She looked me square in my face and said, "Honey, I think I'm gonna die."

The voice of God said, "I've got her, just trust Me."

I started crying and she just fell in my arms. I started CPR on her. My daughter Jena was downstairs. So I yelled for my baby girl downstairs to call 911. The police and ambulance arrived and we went to Fairbanks Memorial Hospital first, then she was medevacked to Anchorage, Alaska.

In the midst of all of this I heard God say, "Will you follow Me?" **Matthew 4:19.**

I looked around and I couldn't see anything near me, but because I've grown up in the church, I knew it was the Lord. He kept saying, "I've got her." Jeanette and I were in the medevac, and the kids and the family members came later.

We went through the whole process of going to the Fairbanks Memorial Hospital to check in. The doctors eventually told me what had happened. Jeanette had an aneurysm and it popped in the back of her neck. God continued to talk to me in spite of it all. The ER doctor wanted Jeanette to be transferred to Anchorage Hospital. As I was riding right beside her in the medevac aircraft, I was praying. The doctors told me that they had put her on the life support equipment. One of them explained to me that he didn't think that she would make it and if she did, she would be a vegetable and no longer the woman we knew and loved. The family and other people were gathered around us. People were praying and making phone calls of love and other communication to support us through this terrible time. The doctor came in and continued to give us an updates. They also showed us the life support equipment and said that if they pulled the plug, everything would stop and she wouldn't be able to breathe on her own... she would be gone. Jeanette held on for a few days until her mother came into town. There was a lot of sadness, confusion, fear, and concern about what was next. That was one of the toughest seasons of my life. After a few days and there was no progress in Jeanette's health, the kids, her mother from Utah, and the rest of the family all came in to say their goodbyes. Because Jeanette was an organ donor, they were able to donate her heart, liver, and kidneys to others and save their lives. To this day, there is someone else walking around today living because of Jeanette.

It was a sad day, one of the saddest days of my life. They pulled the plug; she didn't respond, and she passed away in 1998. God had clearly gotten my attention after that. My 23 years of marriage ended with "'til death do us part."

Our family, including my wife, Jeanette, had been well known in the community. People knew us and came from everywhere to support us, from the church plus parties, and weddings that I deejayed. I also had my Masonic family that helped me through this terrible and sudden loss. I am still, to this day, amazed by the love and generosity shown to me and my family, especially by making travel arrangement for my families from North Carolina and Utah to attend the funeral. I had two clients fly my mother-in-law and her nine sisters from Utah. My family came in from North Carolina, along with anybody else who needed to get to Fairbanks for the funeral. It felt like the whole city had come through to show their love and support during this sad and terrible time in all our lives. A lot of people didn't even know about my hidden, dark life...drugs, drinking, and partying, because I thought I was hiding it well. I wanted to ease the pain by partying even harder. Everybody was shocked that Jeanette was gone, most of all me. I love my family. I never stopped loving them and appreciated all that my family did to help and support me during one of the lowest times of my life.

New Life

How do you move forward and pick up the pieces of your life when you feel like the glue that was holding it together is gone? I've learned that it takes God and taking it one day at a time. After the funeral and everybody went home, my youngest sister, Celestine, stayed with me for a while. She stayed with me to kind of keep me going. I was still trying to work at the radio station. I felt like a walking zombie, just trying to get up every morning and make it through the day.

The kids were trying to move forward with their lives, after their mother was gone. Trina finished college and moved to Atlanta. Jena was going to college and moved out of the house with her sister to Atlanta. James had already moved out of the house because he got married and was living in Anchorage. My youngest, Jena, was about 18 years old when all of this happened. No matter what their ages, my kids will always be my babies. The house was emptying out. Leslie was nearby but still in her own house with her one son, in Fairbanks.

I was there at the house by myself mostly. I got a little money from the death benefit but I was still doing my thing. God had gotten my attention, but I still was not willing to surrender my life to Him. Not yet!

Sometimes it takes others to point you in the right direction and get you on track. I remember my friend Jack Townsend came by to get us out of the house, and we went to church. The church prayed for us. **The scripture says, "the prayers of the righteous availeth much." James 5:16** . After their prayers, I started getting better and better. I was traveling back and forth to Anchorage, just trying to find my way. In 1999, I went to my 30th year class reunion back in North Carolina. It was the July 4th weekend. I also went to church while I was home. The message that Sunday was "If it hadn't been for the Lord by my side, where would I be?" There was a guy videoing the service that day. My friend, Phyllis Joyner, who gave the message, had an altar call or call to discipleship. I ran to the altar and gave my life to Christ. That day, right then, I finally surrendered my all to Christ. **Acts 9:17-19.**

CHAPTER 10

I Surrender All

After I gave my life to Christ, just as the scripture says, I became a new creature in Christ. "Old things pass away and behold all things become new." **2 Corinthians 5:17** . Right? It was like a full circle surrender and ironically or, as God would have it, because of my 30th class reunion in North Carolina and at my home church. Even at the reunion, people saw something. They were looking and watching me, because I was a broken vessel. I was hurt from the loss of my wife and trying to find my way to somehow heal. But if it hadn't been for the Lord by my side, where would I be? I gave my life to Christ that very day.

Back in Alaska, I had a girlfriend, who is my wife today. Prior to going to the class reunion, she took me to the doctor because I had had congestive heart failure. I was in the hospital three times in less than one year. The doctors couldn't seem to get control of my blood pressure. That would have stabilized my heart, but I was still doing cocaine, before I surrendered to the

Lord. My doctor had told my daughter, "If your father doesn't change his ways and get ahold of himself and his life, he's going to die." This was before and really why I went to my class reunion to get away and get myself together.

Even my friends had tried to talk to me because they knew what I was doing. My kids sat me down and said, "Father, we know what you're doing." I thought I was hiding it from everybody but I was fooling myself.

So the class reunion's significance was to change my life and, ultimately, it led to giving my life to Christ. It also changed my life and every else's life around me. God would continue to use me to be a vessel for Him. **Jeremiah 18:2-6.** He had taken me on a journey that would really help me to heal. I'm still healing by love and forgiveness from the Lord even unto this day.

My New Journey

When I came back to Alaska from the class reunion, I started on my new journey. I got better and stopped doing drugs, went to the doctor, and stopped so much partying. I had a job at a restaurant doing karaoke. Mind you, I was a new creature in Christ; old things had passed away and all things became new. I wanted to learn about the Lord. I told the owner of the restaurant, "I'm leaving this business. I'm going to sell you all of my equipment."

He said, "What are you going to do? What are you going to do?"

I said, "I'm going to church for Bible study on Wednesday night so I can learn about the Lord. I need to know more about Him."

He said, "You are kidding?"

I said, "Naw."

I didn't want to go inside the bar area of the restaurant. I would stay outside. He paid me and I rented my equipment to him. Every Wednesday night, I was going to Bible study at True Victory Baptist Church in North Pole, Alaska. I was learning about the Lord.

As I learned about the Lord, I got stronger and stronger. Of course, my friends were laughing at me because I was on fire for the Lord. I said, "Lord, You saved me. You saved me, so You can save my friends." It was amazing, going to church and getting stronger.

Matthew 28:19. I called my father each and every Saturday morning. This one Saturday when I woke up, my father was on my mind. I called him and asked, "Father, are you saved?"

He said, "I don't know."

I said, "Let's make sure." Even though we were on the phone, I got down on my knees and prayed the Sinner's Prayer with my father. It was approximately a year later that my father passed away. I was honored to share the gospel and lead my father to Christ.

I became a deacon in the church, getting stronger and stronger. I was still deejaying, though. Somehow I didn't want to quit. I thought it was all right to be paid to be a DJ because I was striving to take care of my household. Louise and I hadn't really committed to each other. We would eventually get married because God told me he was going to give me another wife. He said, "You've got to take good care of this next wife that I am going to give you." He said, "Love thy wife as I have loved the church and gave myself for it." **Ephesians 5:25-33.** That was Biblical. I knew it was God; I didn't know these things. He started revealing things to me. I sold my house in North Pole and moved in with Louise in Fairbanks. My kids were on their journey and I was on my journey, too. We had to buy a house, a big house up in the mountains, where we lived for a while. After we started attending church regularly, a deacon at the church said to us, "When are you going marry that lady? You are missing out on your blessing." She was talking about me and Louise.

Prior to getting married, I had a lot of work to do to straighten out my life. It was hard to keep my business straight, including my money, my health, and my life, which was so out of balance. My life was out of control, but God was always in control.

When I first met Louise, she was an account executive brought in by the management at the radio station, who called her in to help. She had an appointment with the general manager. I thought she was very attractive. I knew Louise even before my wife Jeanette passed away. Louise had dated one of my best friends, and we all went to a Freddie Jackson concert together.

Time went on, and I actually started a friendship with her. We met and starting walking, down the river path downtown, during the lunch hour. Once she realized the state of my business she always said, "I'm not marrying you until you file your taxes." I had filed for bankruptcy but hadn't filed taxes in eight years. It was a mess. My wife Jeanette had been concerned about my taxes, too, but I paid her no mind. When I finally went to see about my taxes, I owed $155,000, which included all of the fines associated with the taxes. God showed me favor, because when I went to court, the judge told me, "Mr. Buie, pay your taxes." Because I was doing all kinds of good works in the community like fundraisers, etc. I employed my tax accountant, a former IRS agent. He wanted to offer the courts a compromise of $30,000. That was a lot lower than $155,000. I said, offer them $16,000. He said, "Sit down, sit down Mr. Buie. I don't think they're going to take that."

I said, "Offer $16,000." He offered them the $16,000 and they accepted! It was miracle from God. I filed and paid my taxes, and have been paying my taxes on time ever since.

So Louise and I got married, on a riverboat, the *Discovery*. It was a commercial and historic tour boat, that sailed down the Tanana River. I always said if I were to get married again, it would be on that boat. We initially had 400 people invited and knew that we had to narrow the list down. So we decided that we could only invite people we both knew. That shortened the list down to 135 people. So the invitation list was done, and next we had to prepare for the rehearsal and wedding. I wanted a song for Louise to come down the aisle on. Of course, I was deejaying so it should have been easy and I was playing a song, "You Make Me Feel Brand New." I knew that wasn't the song. God told me that! I picked up the TD Jakes CD and put it in my hand, listened to the #7 track. When I put it in the player, the song was "God Gave Me You."

I said, "Louise, Louise, this is the song you're going to come down the aisle on." That song just blessed my socks off. We had a wonderful celebration on the boat. We had nine ministers in attendance. It had been raining, but the skies opened up, and we have pictures of the sun shining down on Louise and me. It was just amazing. We rejoiced and people gave toasts, offered congratulations, and took pictures during the wedding and celebration on the river boat *Discovery*. My children were there, but all of my family couldn't come from North Carolina, and her family was in Germany, but what family and friends were there, we enjoyed the celebration. God gave me a scripture, "Trust in the Lord with all your heart, lean not unto your own understanding, but in all thy ways acknowledge him and He shall direct your paths." **Proverbs 3:5-6 KJV.**

Louise and I started our journey. We were officially one, and we went all the way to Victoria, British Columbia, to continue our celebration and enjoy our honeymoon. **Ephesians 5:31-33.** When the kids and other family and friends came for the celebration, it was really a life-changing moment in 2002.

New Life and New Wife

I was now blessed with a wonderful wife, Louise. She is beautiful, smart, but also an award winning trap shooter. She shot competitively and was the first African-American to represent the state of Alaska at the 100th USTA Grand American World Trap Shooting Competition. She is very talented and goal oriented. She competed a little bit after we were married, but she eventually quit shooting.

She was the advertising director for the *Daily News Miner* newspaper in Alaska, and that's how our paths initially crossed when she came to the radio station. God has blessed us to build a life together serving the Lord.

CHAPTER 11

The Cross Before the Flag

We came back home, and I was now a deacon in my church. I was still going to Bible study and finally stopped deejaying in the club. My love for music would never leave me. I had been in the club before but was now playing the CDs in the church because we didn't have a steady musician at the time. I was really a DJ for the Lord and playing the music that would give glory to God. I also played the drums in church for the music ministry while in Alaska.

I was able to use all of my gifts for God again. It is amazing how my life came full circle, using my gifts for the Kingdom and in the church. **1 Corinthians 13:31.**

The Day that Changed the World and a Vision Was Born

God was preparing me for a vision and work to do, even before I got married. I remember September 11,

2001; I was in bed around four o'clock in the morning... never will forget it. My daughter called me from Atlanta and said, "Daddy, they have flown planes into the Trade Center, turn on the TV."

I thought, "Jena, do you know what time it is?" I turned on the TV and I saw the plane go into the World Trade Center. It was like a movie.

God gave me a song, "Glorify His Name." We had gone to my father's 75th birthday, and my sisters and my daughters wrote this song. God said, "That's why I gave you that song." He started speaking to me and said it again, "That's why I gave you that song."

I said, "Oh my God." By then, I had gotten up, and taken a shower, which actually helped me to wake up. Then I started singing this song. I still know it today. "As sure as the stars are in the sky. His name will always ring on high. He'll always love you. If you glorify His name, you won't be the same." I had that song in my spirit. God said, "That's why I gave you that song." **Exodus 15:1-18.**

I went to work at the radio station, and everybody was in shock. But this day, we started praying. We actually gathered in a circle for the first time and prayed. Some of these people were non-believers.

After that tragedy happened, I ended up at church. The pastor's message was "The Cross before the Flag." I looked around our small church, because we were planning an expansion project in North Pole, Alaska. I'll be honest, I did not see 1.7 million dollar goal needed for the building project right then. Now God spoke to me, "The Cross before the Flag."

I said, "Wow." So the next day, I went down to a graphic artist and told her about the Cross before the Flag and she designed it. We went through a lot of discussion and samples of what "The "Cross before the Flag" should look like and how the Cross has such a great impact on me. My friend even wrote a poem titled "The Cross Before the Flag." The logo was finally designed and that's also when "Together We Stand" was put with the "Cross Before the Flag." Together We Stand was based on **Ephesians 6:14** . "Stand therefore, having your loins girted about with truth, and having on the breastplate of righteousness." The Cross was designed along with "Together We Stand" in a scroll. God instructed me and spoke to me on what to do. I had a lawyer who helped me to get it copyrighted in the Library of Congress in Washington, D.C. I received this vision in stages. I got the Cross before the Flag while I was single and then after I was married God put the "Together We Stand," with it. I didn't get the whole vision together because I was living in sin. But God wouldn't let me let it go. Once I stopped living together with Louise and surrendered my all to Him, He continued to use me and give me His vision. I went to an embroidery business and the lady said, "You need a website."

I said, "A website?" You know, I wasn't savvy on technology or a website.

She said, "Yes, you need 'together we stand.com.'" So, I went looking for people to educate me on how to go about getting a website. There was an organization that told me that a website might cost me thousands or maybe even a million dollars. I prayed about it. This guy all the way in Australia owned the domain name and he agreed to sell it to me for $750. Togetherwestand.com. I said, "Oh my God." So I got the website and then started working in ministry. It was just amazing how God was moving with this project. We created the website, added TWS logo products like t-shirts, hats, leather jackets, mouse pads, pins, and all kinds of items . The website is designed to tell the reader the story. God told me that He wanted me to deliver this vision to the world. I was going up and down the streets, going to different churches telling them all about "The Cross before the Flag" and some people looked at me like I was crazy. I told them that God wants me to deliver this to the world. I said, "Lord, how am I going to do this?" So time went on and one particular morning, I was exercising. I looked up at the television and it was on the Trinity Broadcasting Network.

TBN had a promotion in 2003, "Exalting Him," which was a talent competition. The song that my sisters, daughters and I wrote, "Glorify His Name," had been produced by a cousin with a studio that I had reunited with at a family reunion. I hadn't seen him in 30 years. We went into his studio and he put the song to music. After he put the song together, he called me one night late in April. He said, "Pete," (that's my nickname), "Pete, it's finished." There was an earthquake minutes afterwards. I said, "Lord, what's going on?"

Louise told me, "That was the last word Jesus said, when He gave up his spirit, He said 'It is finished.'"

I said, "Oh, my God. Oh, my God." We entered this song in the Exalting Him 2003 song competition on TBN that year.

There were over 1400 entries and they chose my song to compete in the Atlanta region. As I said, my daughters live in Atlanta. With the help of my cousin, Jon Southerland, daughters, Jena and Trina, my sister, Koku, and four backup singers from Tennessee as well as a band, they practiced and practiced the song. My job was to raise money to help with travel expenses, motel, food, pay the band, and clothing for the televised audition. We all travelled to Hendersonville TN, home of the Trinity Broadcasting Network (TBN) studio for the audition.

There were over 1400 gospel groups and soloists that entered the competition. Our Together We Stand group auditioned in Atlanta, GA and won as a finalist representing the Atlanta region. This was shocking, because everyone we were competing against was so anointed.   Our next stop was auditioning for a worldwide audience at TBN Studio in Hendersonville, Tennessee with the seven other regional finalists.

We didn't win the national televised competition on "TBN Exalting Him 2003." I was amazed that we had a testimony on how the Lord used all of us to bring the song to the world! We gave God the praise and glory. Believe you me, I was blown away how all of this happened in a matter of months.

In 2006, I was exercising and saw on TBN the 100th celebration of the outpouring of the Holy Spirit on Azusa Street in California. I went to the celebration and landed at the Shofar booth and was selected as one of the 12 people to blow the Shofar during the celebration. Oh, what a proud moment, because I had never seen or blown a shofar before. There were more than 50,000 people at that worldwide celebration. I was able to sound the alarm, blow the trumpet (Shofar) for the opening ceremony and worship during that 100-year celebration at Azusa Street in Los Angeles.

CHAPTER 12

Exalting Him 2003, Glorify His Name

After the TBN exposure in 2007, we did a recording, which was our first official music project, the Gospel Festival. Each year, the community and local businesses would help me fund a benefit concert to help the Fairbanks Rescue Mission in Alaska. It is quite expensive to do live concerts. We started doing "Together We Stand" workshop/concerts, bringing the community together. One of the projects completed along with "Together with Stand" was an "America Standing Together for a Miracle of Kindness" This was to encourage and raise awareness to homelessness in Fairbanks and other cities as well. The ministry was coming together. I was asking businesses to help me fund the ministry, to promote and sponsor the concerts. We brought a musical director for the workshop, Minister Norris Garner, to Alaska. He mentioned that I should meet and invite Sheilah Belle of the Belle

Report. She also has a huge electronic magazine and weekly E-blast service. Sheilah has a radio program airing in Richmond, Virginia as well. I invited her to come to Fairbanks, and she accepted to be the Mistress of Ceremonies for the concert in 2008. We tried our best to treat her like royalty when she came to Alaska. To promote the event, I sent out letters and flyers to the churches. Why churches don't like to come together, I don't know, but it happens. We were having auditions for the concert and the day before the auditions we only had one or two groups signed up. Through God and hard work, on the day of the concert, there were 10 choirs, eight vocalists and five praise teams performing. We secured the Carlson Center in Fairbanks, Alaska. For pre-ticket sales, we sold maybe 15 tickets. The next thing I know, the day of the concert, over 2,000 people showed up. It was just amazing to see all these people. We give all honor, glory and praise to God on how He orchestrated the whole thing and brought the people together.

We had everything for a large concert of that size including a big banner, and we donated $5,600 to a the Fairbanks Rescue Mission. We did these concerts and events for several years, and God truly blessed the participants and all those attended the gospel concerts.

We teamed up with Minister Curtis White because he had a production company. We also teamed up with the Church of God in Christ Church in conjunction with their anniversary. Minister White brought Mali Music to be their guest. We had a glorious Holy Spirit-filled concert. Sheilah Belle was there along, with a well-known gospel group called The Barlow Girls. They performed and young girls loved them! Mali Music and his musicians were there to do a five day musical workshop. Mali Music taught some of his original music. He had written a song called "Make Me New," and we paid to record this song live by Mali. He was about to sign a major contract and the fee he charged prior to the contract was a fee that would be a one-time fee. The people told us that after the recording contract, you won't be able to get any music from him at this or any other rate. So we did scramble up the money for the song and paid him for it. He gave us the rights to "Make Me New" and taught it to the workshop. After the Together We Stand Workshop choir sang the song, I told Sheila Belle, while listening to the choir and Mali sing, that this was a hit! Mali sang that song on the Bobby Jones Show, and as a special guest on the American Idol TV competition, as well as numerous live venues. However, the song never did chart on any of the music charts.

God showed us favor and "Make Me New" was on 2012 WoW Gospel CD. This CD featured other top gospel artists in the recording industry. Louise and

I are the executive producers of that song, "Make me New." For it to be accepted on the WoW Gospel CD and not even chart is a miracle. It ended up being a big news release in the paper and it was just amazing how it came to be. To further expose the song, the song and the recording were nominated for a DOVE Award. To God be the Glory!

We had a CD release party at the community park including the Cross before the Flag. The pictures were included in the Belle Report, which was truly the favor of God. Mali Music was singing "Make Me New" all over the USA, which created more attention for the song. Things don't always happen like you want them to happen. If God says that it will happen, some way or somehow the saving, healing, and restoring power of God's message will make it happen. This event happened during the July 4th weekend, as well. There have now been two significant incidences during July 4th. I believe that these things are making me more dependent on God so that He can use me and work through me for His Glory and Praise.

The Cross Before the Flag and the President

For years I had been inspired by President George W. Bush. In 2007, I was at a picnic and met a young woman who was with Public Affairs for Eielson Air Force Base. I gave her my business card and thought nothing else about it. I was a minister so I was ministering everywhere.

Unbeknownst to me, President Bush was scheduled to come to Anchorage, Alaska, but somehow I missed his first visit. Later on in that year, he was scheduled to come back to Eielson Air Force Base. My wife read a news article in the paper over breakfast one morning. I knew I had to make my move so that I could give him the stained glass plaque that I had made for him as well as the pin I created, which was very similar to the flag lapel pin that he wore all the time. He was scheduled to appear at 1:00 p.m. but because of the Secret Service, security, and terrorism, you never know exactly when the President will arrive.

I told my interim pastor, Dr. Garcia, and the head deacon to wear black suits, white shirts, and red ties because we were going to meet the President and I had a presentation to make to him. I told others about it but they laughed and thought I was crazy. But on that scheduled day around 11:00 a.m., the pastor, the deacon and myself headed to the base. At the gate, they asked me what I was there for. I told them that my name wass Anthony W. Buie and I had a presentation for the President on behalf of Together We Stand, American Standing Together for a Miracle of Kindness. I was admitted on the post along with the deacon because we had our military ID, but the pastor had to get a pass to get on post. We were escorted to the command post and the head officer asked my business there. I again told them my name and who I was here on behalf of and about my presentation to the president. They asked to see what was in the briefcase and they said to themselves after seeing it, "He would like that." I got to the Secret Service and told them the same thing. They took my briefcase and said they would get it to him after review. I was disappointed that I didn't get to present the stained glass plaque myself but I was putting my trust in God that it would get there.

I went back outside to wait for the President to come and he didn't arrive until around 5:00 p.m. There was a little old lady who was a part of the Fairbanks Community Center, standing among the more than 10,000 people, who wouldn't have been able to meet or see the President because we were standing in the back of the crowd. I took her by the hand and it looked like the sea of people parted just like at the Red Sea for Moses. The people standing weren't happy, but we went directly to the front on the front row. After his speech, the president came down off of the platform and I stretched forth my hand and said three times, "Mr. President, be blessed."

He replied, "I am blessed." The lady I was with actually got a picture with the President. A few days later, one of the young women who attended our church called me and said that she worked with Public Affairs for the Air Force Base. I had met her previously at the church picnic. She wanted me to know that they took pictures of me shaking the President's hand and I would be provided a photo. So I have a picture of me with the President, and the Cross before the Flag pin and the plaque are sitting on a shelf in the George W. Bush Library. This is perhaps one of the proudest moments of my life and greatest accomplishment to date. After seven years, mission accomplished.

CHAPTER 13

Alaska to Louisville

When I was working at the radio station, the Northwest Broadcasters, I found out that the company would be sold. One of the companies interested in buying the radio station was WLOU. The representative saw what I was doing throughout the community. He told me that if for any reason that I wanted to move to Louisville, Kentucky, that I would have a job. I listened to his pitch and he also took me to lunch. My wife and I took a road trip. We went to North Carolina, Tennessee, Georgia, and Virginia, all as a possibility of relocating and potentially finding our new home if we moved. I was here in Louisville, Kentucky, in 1969 going to school, and although it had been a long time since 1969, I liked it here. Louise's son was here, along with the grandbabies, so we said, "Why not?"

Things were happening in the church and things were happening in my life, and so God seemed to be moving us out. We put our house on the market and we put it in God's hands, and He orchestrated the whole thing. We knew it was God because I had met the man who bought our house at a Martin Luther King celebration. We exchanged email addresses and I contacted him when we put the house on the market. He was a professor at the University of Fairbanks, and he was looking for a home. After contacting me via my email address, he came to look at our house and bought it. And I said, "Wow." He was moving in. I was moving out. We had a big yard sale and then moved and bought a house farther down in the valley of Fairbanks. We moved in there until we were able to leave together. So, we just left. No big party or big celebration, we just left.

Louise left Alaska in Christmas Day, 2011, . She went alone, except for our two dogs. I had to stay behind, because on July 4th I was in a terrible accident. I ran a red light and hit a young father on his motorcycle and he was killed. I could and should have been dead myself, but God again allowed me to live to tell the story. I faced 10 years in prison. It would take an act of God, letters of family and friends defending my honor, and a track record of helping so many people in the community to save me. The grieving wife stood up and told the court that she missed her husband greatly but she forgave me. Just like God forgave me of all of my sins, and I didn't deserve it. I didn't deserve to be forgiven by this young woman either, but GOD and His spirit that reside in her heart gave her the strength to forgive me for a terrible accident. Even the judge said that it could have happened to him as well; it could have been him and that I didn't deserve 10 years in prison for an accident.

After that terrible accident, court, trial, and being in the newspapers, once I was cleared to leave the state, I knew that it was time to leave. God had provided the way for me to move out of there, and the timing was right. I honestly didn't want to leave because I had so many opportunities there in Fairbanks, Alaska, but it was time for us to leave.

We came to Louisville; I had an interview with the station manager and he told me all the great things happening in Louisville. I tried to bring or incorporate what I'd learned in Alaska, but it seemed like I was stepping on so many toes. Other people saw me getting knocked around, not a few but a lot of people did, I didn't get it. So I finally saw the handwriting on the wall and I left. It was time for me to leave, so I resigned from WLOU radio station.

Afterwards, I began working in a sales job with Centronics, a sign company. They hired me after an interview of about five minutes. I found myself on the road, going from Virginia to West Virginia to Tennessee selling signs.

Some time later, I decided to come back to WLOU at their request. I gave them another chance and then

I had to retire. Today I am focused on what God has called me to do. After I retired from the radio station, I didn't really feel right leaving my wife home by herself here in a new area without me. I didn't really know the city that well.

I personally wanted to know the people as well and, more importantly, I was trying to find a church. I tried several churches; some didn't work for us. We did not feel at home or the presence of the Holy Spirit. We moved around a lot, but I was in ministry and wanting to minister to God's people and the world. We brought the National Day of Prayer to a church as well as brought the Cross and the Flag ministry to this same church.

We also had a big event on the Big Four Bridge which goes from Louisville, Kentucky, across the Ohio River to Jeffersonville, Indiana. It was an Easter Sunrise Service, which we also did in Alaska before moving to Louisville. We did that on the bridge for 14 years. It started off with about 20 people and ended up with about a thousand people there on the bridge. We invited Louisville to the Big Four Bridge for the Easter Sunrise Service. It was a great event, but the people didn't show up as I hoped. I'm in this new city, people didn't really know me and didn't respond. God blessed me because of my heart and what I was doing. The company that I paid to have the event gave my money back to me. That was a blessing in itself. Whoever the event was supposed to minister to was ministered to.

We eventually found Canaan Christian Church. where the pastor really preached the Word, and we have been at that church about seven years.

CHAPTER 14

Life Now and Beyond

In 2 Corinthians 5:7, it says that we walk by faith and not by sight. That's clearly the journey that I am on right now. I'm still finding my way and striving each and every day to fulfill the purpose that God has for me on this earth. I have been through so many experiences, trials, tribulations, mistakes, and blessings, but I'm inspired by God to continue until this day. My days are filled with purpose by helping others, spreading the gospel, spreading the mission of the Cross Before the Flag, as well as helping the homeless and veterans in my area to live the best life that they can.

Louisville Rotary Club

Retirement, what is that? I am currently a member of the Louisville Surburban Rotary Club and in fact I am the first African-American president of this club. I was president back in 2016–2017, and now I will hold the position again through 2022. I am proud to support businesses and other organizations in this area to partner, collaborate, educate, and thrive to be the best that they can be.

Winning Souls for Christ

At this stage in my life, I think God is speaking to me more and more about winning souls for Christ. My job, in addition to my other endeavors, is to teach, love, and seek to evangelize His Word so that it can permeate the hearts and transform the lives of many and let them know if God can change me, He can definitely do it for them. I believe that He brought me to Louisville at this time and for this purpose, because here I am on a bigger platform so that I can promote His message of restoration of families, deliverance, and support for the homeless and veterans through my book, the Cross Before the Flag, and the PUSH Initiative (Providing Unfortunate Souls Housing). I am a member of the Coalition for the Homeless to helps to find solutions for this growing problem, especially in our community. I need and want desperately to get this book out before God allows me to come home to be with Him.

NABVETS

Well, I met Commander Shedrick Jones at WLOU. I believe that when we met, I was trying to sell him advertising on the radio station but it just didn't work out. They had an outrageous price and it just wasn't fair. After he met me and we had time to talk, he wanted to know what I was doing there at the radio station. He saw so much talent in me, but I really didn't understand what NABVETs was all about. Commander Jones explained to me that NABVETS stands for the National Association of Black Veterans. We got together outside of the station, talked more, and he said he wanted me to become a lifetime member. So I looked at the organization, understood what they're trying to do, their initiatives and purpose, and I joined. Commander has opened many doors for me by connecting me with the people in the city. We are doing a great work by cleaning up and restoring the local cemetery that has veterans buried there. I said, "Wow, what a great project!" when I heard about it. We have been a part of so many events as well as helping to be an advocate by going to the capital of Kentucky in Frankfort to voice our concerns for veterans and their unique situations and problems. I am an unofficial chaplain for the NABVETS chapter.

My wife and I were included in the Kentucky Monthly magazine for our efforts and it brought more attention to our purpose and mission. So coming from Alaska to Louisville was a part of God's purpose for me and my family, and believe you me, God has me just like the potter has the clay. He is shaping and molding me to be a soldier for Him each and every day. I don't want to stand before Him and have Him say He never knew me. I want to seek each day to make disciples for Christ, wins souls for Him, teach, and evangelize for His Kingdom. I want them to know Him so that He can transform them into what He wants them to be. That's my legacy and inheritance, not only to my family, my kids, and grandkids but to the Body of Christ.

Conclusion and How You Can Help

The main purpose of this book is to show you God's forgiveness and the transformational power of His Word and ways. Secondly, my goal is to encourage everyone who reads this book to keep moving forward and not to look or go back to the ways and things that are not pleasing to God. Finally, I want each of us to strive to keep, maintain, and bring back God into every area of our lives, cities, schools, states, and the nation. My goal even to this day is "Together We Stand, America Standing Together for a Miracle of Kindness" to benefit the homeless, veterans and those unfortunate people throughout the world. **2 Chronicles 5:13-14.**

To support this endeavor, please visit our website, *www.togetherwestand.com*, call us at office cell (907) 460-8820, or email us at *anthonyawbuie@gmail.com*.